ELEMENTS OF THE PERIODIC TABLE

WHAT IS SODIUM?

011

Na

SODIUM

22.99

KATHLEEN A. KLATTE

ROSEN PUBLISHING

Published in 2026 by The Rosen Publishing Group, Inc.
2544 Clinton Street, Buffalo, NY 14224

Portions of this work were originally authored by Mary Thomas and published as *Sodium*. All new material this edition authored by Kathleen A. Katte.

Designer: Rachel Rising
Editor: Kathleen A. Katte

Library of Congress Cataloging-in-Publication Data
Names: Klatte, Kathleen A., author.
Title: What is sodium? / Kathleen A. Klatte.
Description: Buffalo, New York : Rosen Publishing, [2026] | Series: Elements of the periodic table | Includes bibliographical references and index.
Identifiers: LCCN 2024057595 | ISBN 9781499478525 (library binding) | ISBN 9781499478518 (paperback) | ISBN 9781499478532 (ebook)
Subjects: LCSH: Sodium--Juvenile literature. | Chemical elements--Juvenile literature.
Classification: LCC QD181.N2 K557 2026 | DDC 546/.382--dc23/eng/20250212
LC record available at https://lccn.loc.gov/2024057595

Manufactured in the United States of America

CPSIA Compliance Information: Batch #CSRYA26. For further information, contact Rosen Publishing at 1-800-237-9932.

CONTENTS

INTRODUCTION

Salt is a very common material. After all, you probably see it almost every day. It's in the saltshaker on your dining table. It's in the bag of rock salt that may be next to your front door in winter. Roads are treated with salt or brine to make them safe to drive on in icy conditions. There's even salt inside your body. And some people have had the experience of getting a mouthful of salty seawater at the beach. Yuck!

The salt that we're most familiar with is a compound of sodium and chlorine—sodium chloride (NaCl). Sodium (Na) is the most common alkali metal and the sixth most abundant element on Earth. Sodium is extremely reactive—it bonds easily with other elements. For this reason, sodium is generally found in nature as part of a compound. These compounds make up almost 3 percent of Earth's crust.

Sodium chloride occurs in nature as a component of seawater or can be mined from the earth as rock salt. Salt licks are places where brine or seawater evaporated, leaving salt behind. Because animals need salt, game trails would lead to natural salt licks. Ancient humans followed, and these trails eventually became roadways.

Humans have used salt for centuries. The Egyptians used it as far back as 2000 BCE to preserve food. Refrigeration is a fairly new technology, so for most of human history, anything that could be used to keep food safe for eating has been highly prized. In the sixth century CE, Moorish merchants valued salt as highly as gold. In the 13th century, Marco Polo brought back tales from his travels of cultures that used salt coins.

Ancient cultures also recognized the antiseptic properties of salt. Egyptians used a form of salt called natron as part of the process of mummification—preserving dead bodies for their journey to the afterlife. The economies of many major medieval and renaissance cities were based on the salt and spice trades. It's used in many industries, especially the manufacture of dyes and perfumes.

Salt also has special significance in many cultures. Spilling salt often is considered unlucky. This belief probably originated because it wastes something that society considered highly valuable. Until the 18th century, a person's social prominence could often be noted by how close they were seated to the salt at the dining table. For centuries, bread and salt have been offered to guests as a symbol of hospitality and trust. By taking a closer look at the element of sodium, we can better understand why salt is considered so important.

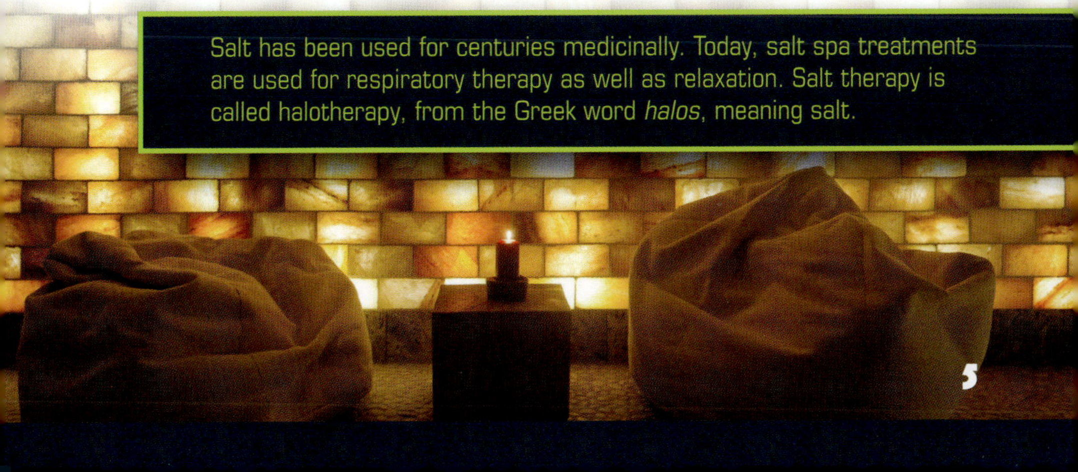

Salt has been used for centuries medicinally. Today, salt spa treatments are used for respiratory therapy as well as relaxation. Salt therapy is called halotherapy, from the Greek word *halos*, meaning salt.

CHAPTER 1

A HISTORY OF SODIUM

Sodium is all around us. It's in the seawater and even in the soil. Plants need small amounts of water-soluble salts, and they get them from minerals present in the soil they grow in. However, too much salt can harm most plants.

Salt is one of the oldest food preservatives known to mankind. Walk through a supermarket and see how many products contain salt. There's all the different kinds of chips and crackers in the snack food aisle. Pickles, sauces, and canned meat products all use salt as a preservative and flavor enhancer. Read the nutrition labels on random products and see what else contains salt.

There's also sodium in space. The sun contains trace amounts of sodium. It's also been found in the atmosphere of Mercury in the form of sodium vapor. Spectrometer analysis carried out by NASA's *Galileo* spacecraft shows a sub-surface saltwater sea on Jupiter's moon Europa. There's even a layer of Earth's atmosphere comprised of sodium. The sodium layer is visible from space as a yellow-orange glow.

HUMPHRY DAVY

Sir Humphry Davy (1778–1829) was on to something. It was 1807, and the chemist and professor had been reading up on the work of two other chemists, Alessandro Volta (1745–1827) and William Nicholson (1753–1815). Davy began by researching the work of Volta, an Italian physicist known for his work in the study of electricity and his invention of the first battery in 1800. Davy also studied the work of Nicholson, an English chemist who discovered the electrolysis of water. The electrolysis of water separates water into its basic elements, hydrogen (H) and oxygen (O). During electrolysis, wires are attached to a battery at one end, while the other ends of the wires are placed in a jar of water. When the electric current hits the water, the hydrogen and oxygen in the water will begin to separate and form bubbles at the submerged ends of the wires. Davy wondered whether electrolysis could be used to separate potash—the ash left after wood burns—into its basic elements.

Some of the world's salt is still harvested through traditional means, such as evaporation from salt lakes.

The industrial drilling equipment in this salt potash mine looks like something out of a sci-fi movie.

To conduct his experiment, Davy built his own electrolysis equipment, consisting of two lead bars called electrodes, each attached at one end with wire to a battery. After several false starts, small silvery drops formed on one of the electrodes. This metal was like nothing else Davy had seen before. It was soft and extremely lightweight. Davy had succeeded in isolating the basic elements of potash! He called this new metal "potassium," derived from the medieval Latin word *potassa*.

Davy was determined to discover what difference, if any, there was between potash and soda ash, a harsh substance that remained when wood ashes were boiled in water. Davy tried the electrolysis experiment once more using soda ash. The experiment helped to break down the soda ash into hydrogen, oxygen, and another strange silvery metal. It, too, was very soft and lightweight. The metal bubbled and fizzed whenever it was in contact with water, although it floated on the water's surface.

Exposed to air for a long period of time, this new substance started to look a lot like the salt used to flavor food. It even tasted like salt. Davy called this second substance "sodium," from the medieval Latin word *sodanum*, which referred to a headache remedy. Little did Davy know he had discovered one of the essential ingredients for life on Earth.

ELEMENTAL SODIUM

Just what is an element anyway? Elements are only one kind of substance. They are the building blocks of matter, which you can think of as just about anything and everything that takes up space. Try thinking about matter as a giant puzzle and elements as the puzzle pieces—really, really small puzzle pieces. By fitting one "piece" of sodium (Na) together with one "piece" of chlorine (Cl), you can create a molecule of salt (NaCl). Different combinations of elements create everything in the universe, from plants to plancts.

Himalayan salt lamps are a decorating and health fad. The rosy glow can be soothing, and many people believe the salt helps purify the air. However, there are no scientific studies supporting these claims. The lamps also pose a danger to pets, especially cats, who can ingest toxic amounts of sodium by licking the chunk of pink salt.

Let's talk a little more about these puzzle pieces and how they fit together. For starters, a more proper name for them would be molecules. If you've ever put together a puzzle of your own, you know that even puzzle pieces must be made of something, like cardboard, plastic, wood, or—in the case of chemistry—atoms. An atom is the smallest piece of an element that can join with another particle to form a molecule. An atom of an element cannot be broken down any further. However, there are even smaller pieces that make up an atom, and these are called subatomic particles.

WHAT'S SMALLER THAN AN ATOM?

Atoms themselves are made up of tiny particles called protons, neutrons, and electrons. These are known as subatomic particles. Protons have a positive electric charge, electrons have a negative electric charge, and neutrons have no charge at all.

If you were to view an atom through a very powerful microscope, you would find that it is largely made up of empty space. The rest consists of a positively charged nucleus, which is the center of the atom. Inside the nucleus are protons and neutrons. Surrounding the nucleus are shells of negatively charged electrons. An element's atomic number usually is the number of protons contained within its nucleus. Sodium has 11 protons in its nucleus. Therefore, its atomic number is 11. Most sodium atoms have 12 neutrons.

The nucleus of a sodium atom is small and heavy compared to the surrounding electrons, which are the lightest charged particles in the universe. If you've ever experimented with magnets, then you probably know that opposites attract—negative and positive charges are drawn toward each other. In an atom, these forces, called electromagnetic forces, bind the electrons to the nucleus and help hold the atom together.

Outside the nucleus are the shells that contain the electrons. The first shell can hold only two electrons, the second shell can hold up to eight electrons, and the third shell contains up to 18. The behavior of an atom is strongly influenced by the electron distribution in shells. The sodium atom has three shells of electrons. The inner shell contains two, the second shell holds eight, and the third is home to only one electron.

At room temperature, pure sodium is a soft, silver-colored metal that can be easily cut.

SODIUM

CHEMICAL SYMBOL: Na

PROPERTIES: Soft, silvery white metal that floats on water and combines easily with many other elements

DISCOVERED BY: Element has been known of since prehistoric times; isolated by English scientist Sir Humphry Davy in 1807

ATOMIC NUMBER: 11 **ATOMIC WEIGHT:** 22.9 amu

PROTONS: 11 **ELECTRONS:** 11 **NEUTRONS:** 12

DENSITY @ 293 K: 0.971 g/cm3

MELTING POINT: 208°F; 97.81°C; 371 K

BOILING POINT: 1,621°F; 882.9°C; 1,156.1 K

COMMONLY FOUND: Earth's atmosphere, soil, crust, and oceans, as well as plant and animal life

Sodium's periodic table entry has lots of useful information. It includes its common name, chemical symbol, atomic number, and atomic weight.

Death Valley is the lowest point in the Western Hemisphere. It's below sea level and the air is very dry, evaporating water and leaving salt flats behind.

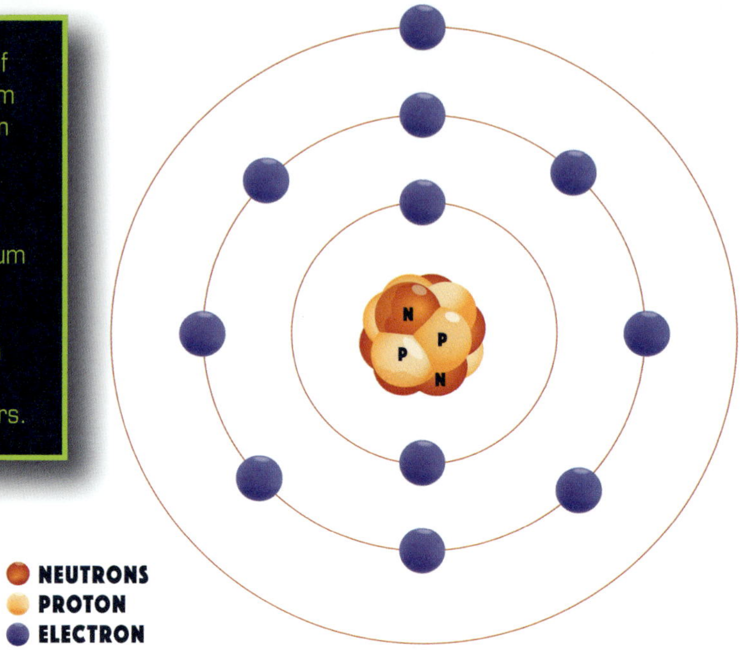

The structure of the sodium atom allows it to form many different compounds. Beyond basic table salt, sodium compounds are used for everything from medicine to fire extinguishers.

THE SEARCH FOR PURE SODIUM

Pure sodium doesn't occur in nature. Because of the way its atom is formed—with one electron in its outer shell—a sodium atom is very reactive. That is, it's always looking for other atoms to share that electron with. The fewer electrons an atom has in its outer shell, the more reactive or unstable it is. This means it combines very easily with other elements to form compounds. Once sodium forms a compound with another substance, its inner shell (which is full) becomes its outer shell. When there is no more space for additional electrons in its outer shell, the molecule becomes stable. This means it can't combine with any additional atoms.

The only way to isolate pure sodium is to use electrolysis to break the bonds between sodium and whatever substance it's formed a compound with. Pure sodium must be kept in an inert, or nonreactive, substance such as nitrogen (N). Otherwise, it would just form new compounds with what is available. This isolation isn't a process that occurs in nature. The pure metallic form of sodium must be manufactured.

When mixed with water (H_2O), sodium will bubble and fizz. This is caused by the reaction between sodium and the oxygen in water as it forms sodium oxide, releasing pure hydrogen gas.

Sea salt is produce when sea or salt lake water evaporates, leaving salt crystals behind.

CHAPTER 2

A PLACE AT THE TABLE

Today, a chart called the periodic table of elements appears in chemistry classrooms and textbooks around the world. It lists the 118 chemical elements currently known. Each element is listed by its chemical symbol, atomic weight, and atomic number. An organization called the International Union of Pure and Applied Chemistry (IUPAC) publishes the table and oversees the process for adding newly discovered elements.

However, this wasn't always the case. In the mid-19th century, many new elements were being discovered, but there wasn't really a good way to organize them. A Russian chemist named Dmitri Mendeleev (1834–1907) noticed similarities between certain elements. He arranged the 63 elements known at the time on a table in order of their atomic weight, from the lightest on the left to the heaviest on the right. The first version of his table was published in 1869.

Mendeleev's arrangement was a simple idea that would help revolutionize the way scientists understood chemistry. The table allowed scientists to really see for the first time the relationships, trends, and patterns among elements. Mendeleev's table had gaps, but he predicted that these gaps would soon be filled by elements not yet discovered. He also guessed that these new elements would have properties that could be predicted based on their placement in what he called the periodic table of the elements. He was quite right, as it turns out. Within 20 years after he first created the periodic table, three new elements would be discovered, and they each possessed the properties that Mendeleev had predicted.

Although the table was not widely accepted at first, scientists eventually realized that it was a pretty good idea. In fact, the modern periodic table is very similar to Mendeleev's original design. Today, the periodic table is arranged in horizontal rows called periods. In each period, the elements are arranged in the order of their atomic numbers.

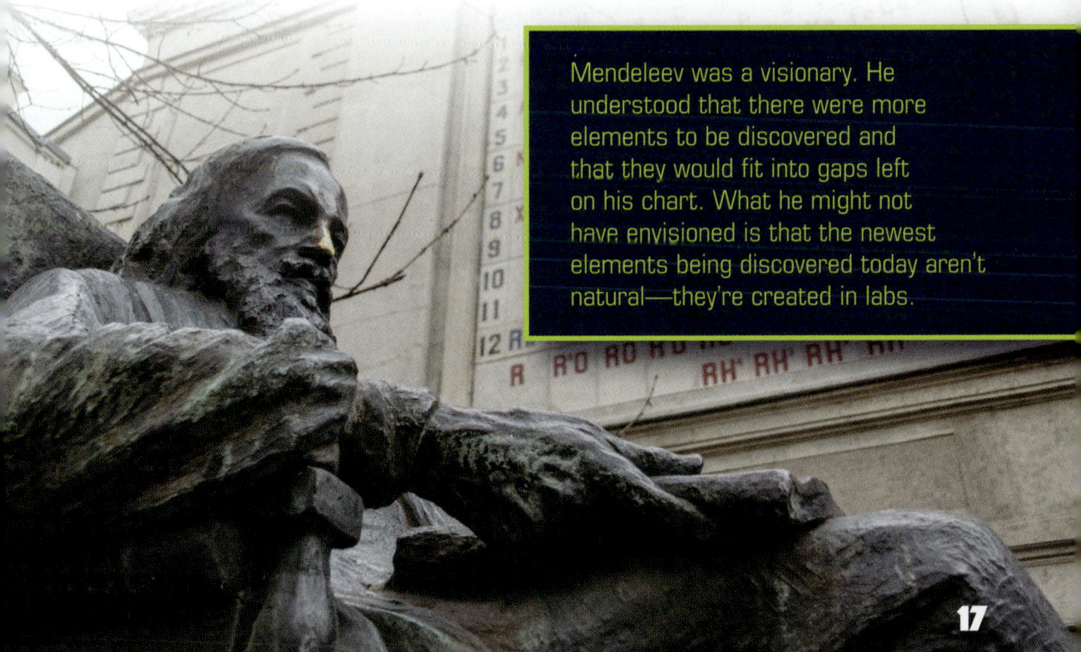

Mendeleev was a visionary. He understood that there were more elements to be discovered and that they would fit into gaps left on his chart. What he might not have envisioned is that the newest elements being discovered today aren't natural—they're created in labs.

THE LATEST ADDITIONS

Mendeleev realized that there were probably more elements in the world than had been discovered. He left spaces in the periodic table for new elements to be added. His theory was proven correct as more elements were discovered and added to the table during his lifetime. Since the publication of the original table, improvements in technology have led to the discovery or creation of more than 50 additional elements. New discoveries are still being made today.

The latest additions to the periodic table are four superheavy elements created in laboratories. In 2016, elements 113, 115, 117, and 118 were added to the table.

ATOMIC NUMBER	INITIAL DESIGNATION	OFFICIAL NAME
113	UNUNTRIUM (Uut)	NIHONIUM (Nh)
115	UNUNPENTIUM (Uup)	MOSCOVIUM (Mc)
117	UNUNSEPTIUM (Uus)	TENNESSINE (Ts)
118	UNUNOCTIUM (Uuo)	OGANESSON (Og)

The elements are also organized into 18 vertical groups. The number of the group appears above each column of the table. In the same way that members of a family often resemble one another, elements within these groups also have similar chemical properties. In fact, they are often referred to as families of elements. By arranging the elements this way, scientists could predict whether any given element was a metal, a nonmetal, or a metalloid, which is an element that has the properties of both a metal and a nonmetal.

The sodium and chlorine atoms that form common table salt (NaCl) molecules are formed by ionic bonds. This type of bond involves a positively charged sodium ion bonding with a negatively charged chlorine ion.

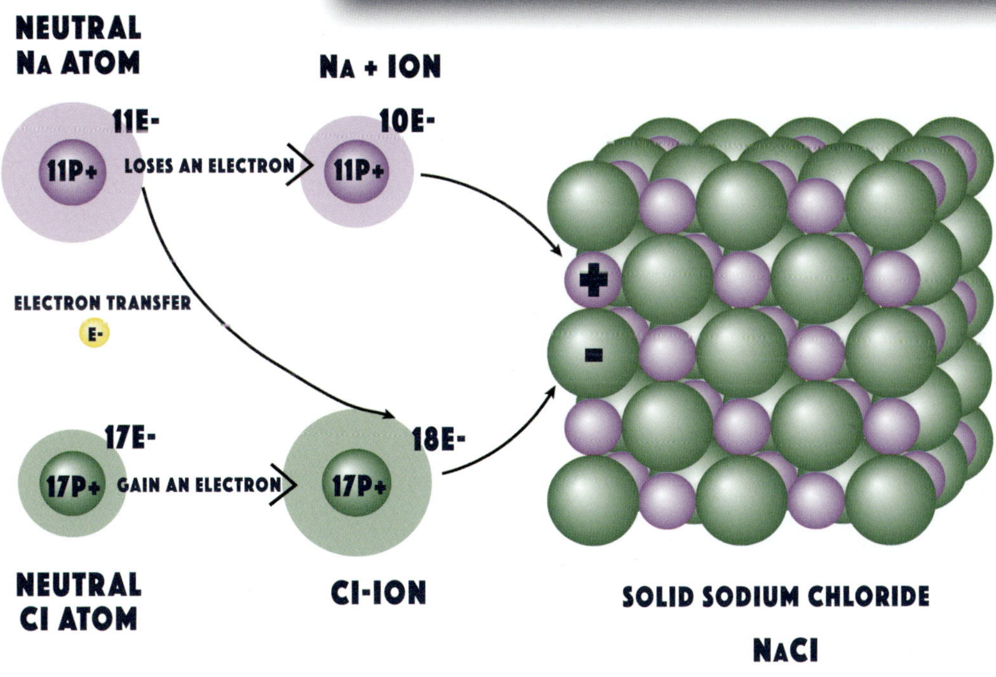

Salt, like this halite, forms crystals because of the way the molecules interlock. The latticed shape occurs naturally at the molecular level and continues up to the scale that can be seen by the naked eye.

A PLACE FOR EVERYTHING

Now, let's take a closer look at the periodic table. Sodium is located within group IA on the periodic table, which also contains lithium (Li), potassium (K), rubidium (Rb), cesium (Cs), and francium (Fr). These elements are known as the alkali metals. Alkali metals are very reactive metals that do not occur freely in their pure form. These metals each have only one electron in their outer shell, and they very often give it away or share it with other elements to become more stable. The elements within group IA behave more similarly to each other than elements in any other group of the periodic table. As you move down the group from lithium to francium, each element reacts more easily with other elements. In other words, the elements grow more reactive. Hydrogen is often placed in group IA on the table because it, too, has only one electron in its outer shell. However, it isn't an alkali metal but a gas.

SODIUM'S ELEMENTAL PROPERTIES

Properties are the characteristics of an element. They are what make the element unique. For example, in humans, your eye color, nose shape, and hair color may be properties that help tell you apart from other members of your family.

Each of the group IA elements, including sodium, are silver metals. They are soft and can be easily cut with a knife to expose a shiny surface that becomes dull when air hits it. As with all metals, the alkali metals are good conductors of heat and electricity.

All elements on the periodic table can appear in one of three physical states or phases: solid, liquid, or gas. Sodium is a solid at room temperature. However, when heated to 208°F (97.81°C), it melts into a thick liquid. If you continue to heat it, sodium will boil into a gas at 1,621°F (882.9°C).

ALL IN THE FAMILY

Most sodium atoms have 11 electrons, 12 neutrons, and 11 protons. These atoms form a stable molecule—that is, one that does not react or form bonds with other elements. Scientists refer to this kind of sodium molecule as sodium-23. It's the only kind of sodium that occurs naturally.

However, some sodium atoms have a different number of neutrons. These are called isotopes. An isotope is an atom of an element which contains a different number of neutrons. Most elements have isotopes. Isotopes are very similar but may have slightly different characteristics. There are about 20 known isotopes of sodium. All are radioactive, and all were created in labs. The sodium isotopes with the longest half-lives are sodium-22 (with 11 neutrons) and sodium-24 (with 13 neutrons).

Radioactive isotopes of sodium are used to find leaks in pipelines.

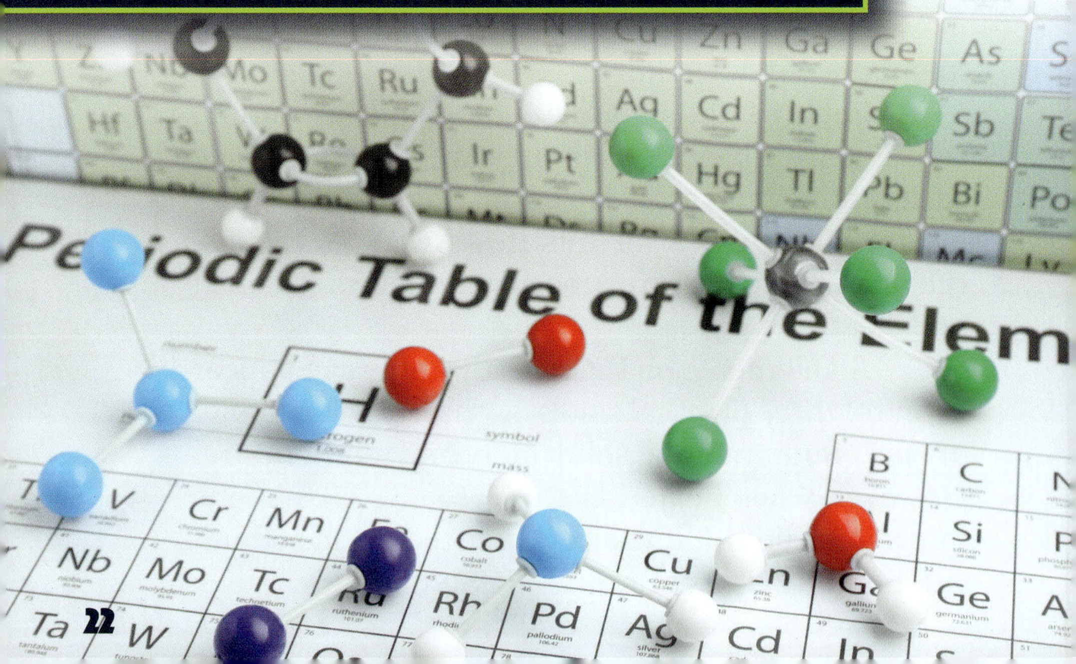

Sodium's chemical symbol Na comes from the Latin word *natrium*.

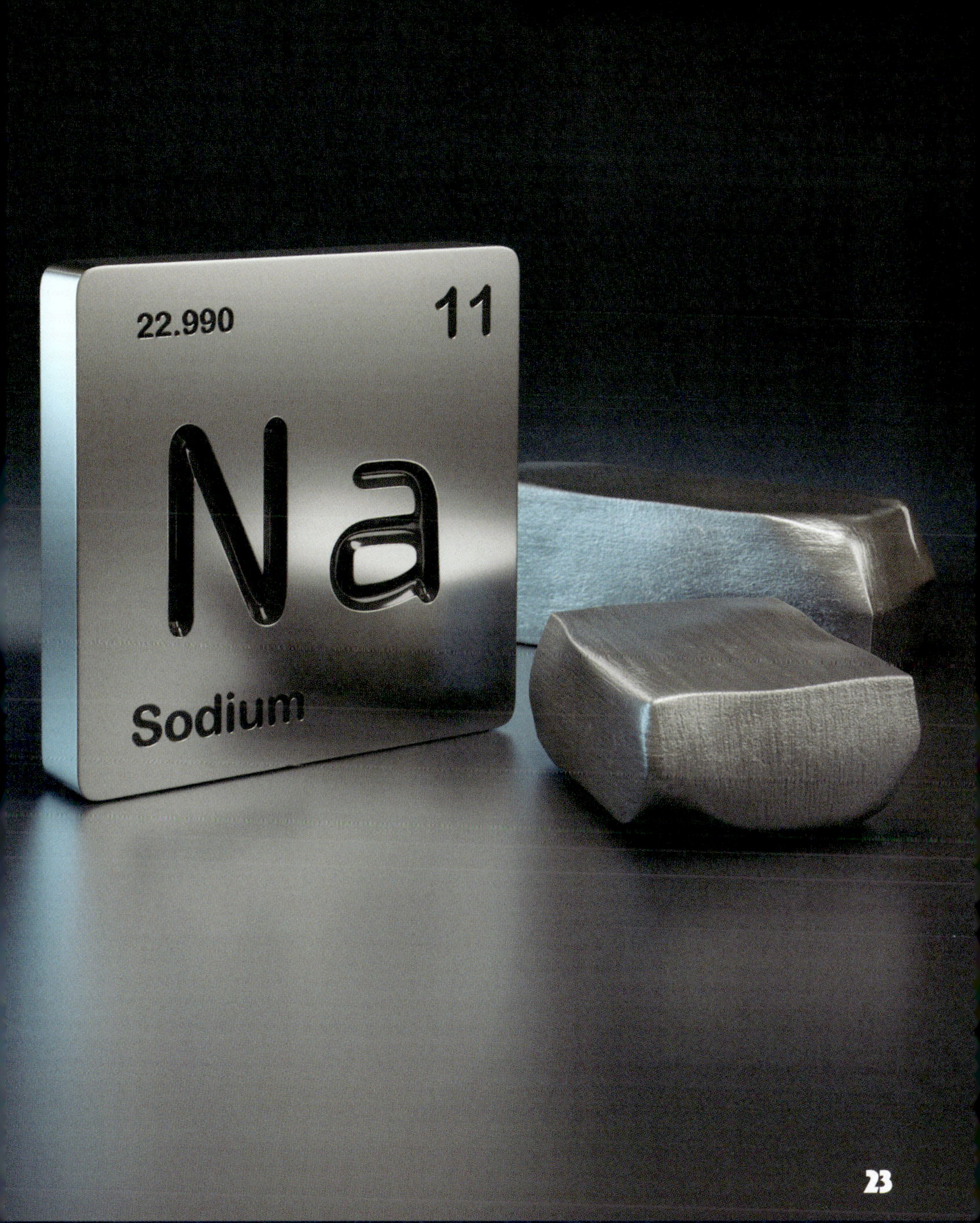

CHAPTER 3

SALT IN THE NATURAL WORLD

Anyone who's ever gotten a mouthful of seawater while swimming at the ocean shore can tell you that seawater is very salty. Seawater is about 35 parts per thousand of salt—much too salty for people to drink. There are about 120 million tons (109 million mt) of salt per one cubic mile (4.2 cubic km) of seawater. This means that most of the world's water is unfit to drink or irrigate crops. But how does it get that way? Where does the salt come from?

A lot of salt comes from the land. Rain is slightly acidic. As rain falls, the acid dissolves minerals from the rocks and soil. This includes sodium oxide, formed when sodium ions react with oxygen in the soil. These minerals interact with each other in the water, forming compounds. For example, sodium oxide combines with chlorine to form sodium chloride—common salt. Rainwater flows over the land, collecting minerals as it runs off into the sea. Some of these minerals are used by organisms that live in the water. Others—including salt—aren't used as much, so they accumulate in the seawater.

Salt also comes from the ocean floor. Some is released by the eruption of underwater volcanoes. Some salt comes from hydrothermal vents. These are cracks in the ocean floor where magma from the earth's core heats seawater. The extreme heat causes chemical reactions that add minerals to the water.

The salinity, or amount of salt in the water, can vary by location or season. In places with warmer temperatures, water evaporates quickly, causing the salt content of the sea to become more concentrated. In places where freshwater rivers drain into the sea, the salt content will be lower. As icebergs melt, they release fresh water into the sea, so cooler seas tend to have less salt.

Some of the salt in seawater comes from salt domes. These are geological features that form over eons. Salt domes form in places that used to be warm, shallow seas thousands of years ago. The water evaporated, leaving behind a layer of salt. Over time, more sediment piled on top of the salt. Eventually, pressure and heat in the earth's crust push the salt to the top.

Aside from making food tastier, salt is also one of the oldest known forms of food preservation.

SO MANY SALTS

The common salt found on french fries and pretzels is only one kind of salt used today. In chemistry, a salt is a type of compound created when a substance called a base, or alkaline substance, is neutralized by another substance called an acid. When an acid is neutralized, the number of ions that it has is completely balanced by those of another substance, so it does not continue to react chemically with its environment. Acids are substances that form hydrogen ions when dissolved in water. They usually have a bit of a sour taste, like grapefruit or lemon juice, for example. Bases easily accept the hydrogen ions created by an acid. They also usually have a bitter taste.

WHERE DOES IT COME FROM?

A large amount of the salt that eventually shows up on your dining table is found naturally in deposits around the world. Some of these are thousands of feet thick, while still others are the remains of ancient salt lakes like Utah's Great Salt Lake and the Dead Sea in Israel. In the United States, nearly 85 percent of what becomes table salt is mined from deep within the earth, where salt deposits formed as ancient lakes and oceans evaporated over millions of years. Some layers of salt are found so far beneath the earth's surface that they cannot be reached with mining equipment. In other parts of the world, water from rain or the runoff from rivers and lakes has penetrated salt deposits but has not been able to

The Great Salt Lake in Utah is the largest saltwater lake in the Western Hemisphere. It's actually much saltier than the ocean.

evaporate, thus forming an underground brine (salty water) well. Thick and sludgy, brine is the substance that Alessandro Volta used to wet the cloth he used in the construction of the voltaic pile, a series of several metal discs each separated by a damp cloth to create a continuous current. The brine from a brine well can be pumped up to the surface of the ground, where the water in the brine evaporates, leaving salt crystals behind.

In a similar process, salt water from the ocean is pumped into large ponds, where it is allowed to evaporate, leaving salt crystals on the bottom of the pond. As the seawater evaporates, different chemicals in the water crystallize into minerals at different times. In this way, scientists and industry workers can mine salts of bromine (Br), a gas, and magnesium (Mg), a metal, as well as sodium chloride.

A molecule of common table salt—sodium chloride (NaCl)— is a compound formed of one sodium ion and one chlorine ion. The sodium ion donates its single outermost electron, forming a strong electrostatic bond. This forms a stable salt molecule.

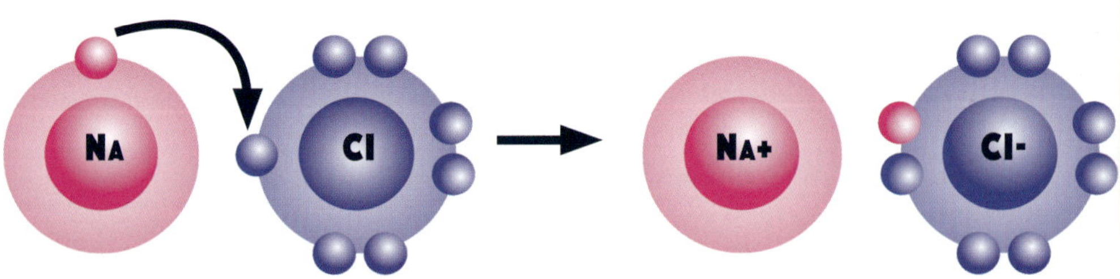

The Dead Sea is a landlocked salt lake located near Israel. It's the lowest body of water on the earth's surface.

SALTWATER INTRUSION

An aquifer is an underground layer of rock where groundwater has accumulated. Wells tap into aquifers to bring fresh water to the surface. Most drinking water comes from aquifers.

Saltwater intrusion is a phenomenon that occurs in coastal areas. Normally, ground water flows toward the sea. This creates a buffer zone, pushing salt water out to sea and keeping the groundwater fresh. However, when too much water is removed from an aquifer, such as by pumping too much from a well, the flow reverses. Salt water encroaches on the aquifer. If the water becomes too salty, it's not useable for drinking or irrigating crops. Groundwater in coastal areas needs to be carefully managed to ensure there's a suitable supply of fresh water.

The Red Sea is one of the warmest and saltiest bodies of water on Earth. Despite its name, the water is generally a brilliant blue. However, it is subject to occasional algae blooms. When the algae die off, the water is discolored with a distinctive reddish-brown tone.

Seawater is usually about 3.5 percent salt. The Great Salt Lake ranges from about 5 percent to 27 percent salt in different parts of the lake. The Dead Sea is about 34 percent salt.

Excessive salinity in soil can greatly impact food production.

SALT IN THE SOIL

There's sodium in the soil too. Most plants need a small amount of sodium to carry out their life processes. The sodium naturally found in soil is generally all that's needed. Rainwater contains trace amounts of salts. More is added by the erosion of rocks.

However, too much of a good thing can be detrimental. Too much salt prevents plants from taking in the water they need to live. If farmers irrigate their crops with water that's been subject to saltwater intrusion, salt can build up in the soil. Fertilizers, pesticides, and other soil additives intended to aid plant growth can also lead to excess sodium in the soil if they're not used carefully.

Climate change and human activity are responsible for increasing soil salinity.

CHAPTER 4

WAYS PEOPLE USE SODIUM

Humans have used sodium for thousands of years to season and preserve food. We all know how a sprinkle of salt improves the flavor of many foods. Ham is made by salting pork, and pickles are made by immersing vegetables in brine. Salt is used to cure cheese. It's one of the oldest methods of preserving food. Salting fish was a major industry in ancient Mediterranean cultures. Salt made it possible to store food for winter or for long sea voyages before the invention of refrigeration. But how does it work?

Food spoils due to the presence of mold and bacteria. These microorganisms thrive on moisture. Salt dries out moisture, killing the mold and bacteria. Salt aids in a natural process called osmosis. This is a process where water moves from an area of higher concentration to an area of lower concentration. If you picture two cells side by side, water will move across the cell membrane from the cell with more water to the cell with less water. Salt speeds up the process of osmosis, drying out and killing bacteria before it has time to spoil food.

When most people think of pickles, they tend to think of pickled green cucumbers. However, before refrigeration became common, people would pickle any vegetables left in their garden at the end of the season.

GETTING CLEAN WITH SODIUM

Sodium is also an important part of soap. In any body of water, each water molecule is surrounded and attracted to other water molecules. However, at the surface of water, the water molecules are surrounded by other water molecules only on one side. On the other side, air molecules touch the water molecules, creating tension on the surface of the water. This surface tension causes water to bead up on surfaces like glass or fabric, which slows the cleaning process by slowing down how long it takes something to become wet. You can see surface tension at work by placing a drop of water on a countertop. The drop will hold its shape and will not spread.

For soap to clean, surface tension must be reduced so water can spread and wet surfaces. Chemicals that can do this effectively are called surface-active agents, or surfactants. They are said to make water "wetter." Surfactants perform other important functions in cleaning, such as loosening, emulsifying (dispersing in water), and trapping the dirt outside of the substance until it can be rinsed away. Soap, detergent, and other chemicals used in cleaning are surfactants.

Soaps are sodium or potassium salts of fatty acids that dissolve in water. They are made by treating fatty acids chemically with a strong chemical salt of an alkali metal such as sodium or potassium that dissolves in water. Fatty acids are weak acids composed of two parts: a carboxylic acid group molecule consisting of one carbon atom, two oxygen atoms, and one hydrogen atom, plus a hydrocarbon chain attached to the carboxylic acid group. Generally, they are made up of a long straight chain of carbon atoms each carrying two hydrogen atoms.

Many years ago, the alkalis used in making soap were obtained from the ashes of burned wood and were mixed with oil and fat. However, during the late 18th century, a French chemist named Nicolas Leblanc developed a way to produce sodium hydroxide (NaOH) and sodium carbonate (Na_2CO_3) out of table salt. During the Leblanc process, table salt was first converted into a compound called sodium sulfate (Na_2SO_4) by mixing it with sulfuric acid. The sodium sulfate was then roasted with chalk or limestone ($CaCO_3$) and coal to produce a mixture of sodium carbonate and sulfur. The carbonate was filtered out with water, and the solution crystallized. The Leblanc process was adopted throughout Europe within just a few years. However, over the past century, many soap-making processes have come and gone, with each process rendering the previous one obsolete.

Soap doesn't just remove dirt from our hands—it also washes away germs that can spread disease.

MAKING SOAP

There are two major ways to make soap today. Saponification of fats and oils is the most widely used soap-making process. This method involves heating fats and oils and combining them with a liquid alkali, such as sodium hydroxide, to produce soap, water, and an oily substance called glycerol. Many years ago, people used the glycerol that was produced during soap making as a moisturizing lotion for their skin. The other major soap-making process is the neutralization of fatty acids with an alkali. Fats and oils are split by using high-pressure steam to yield fatty acids and glycerol. The fatty acids are mixed with and neutralized by an alkali to produce soap and water. When the alkali is sodium hydroxide, a sodium soap is formed.

Soap molecules form a structure called a micelle, surrounding dirt particles and separating them from soiled surfaces so they can be rinsed away by water.

Bubbles form when soap molecules push water molecules apart, trapping air.

HOW DOES SOAP WORK?

Soap is great for cleaning because of its ability to act as an emulsifying agent. An emulsifier is a chemical that can suspend oil and dirt in such a way that they can be removed with water. Each soap molecule has a long hydrocarbon chain, sometimes called its "tail," with a carboxylic "head." The head of a soap molecule is made up of a positively charged sodium ion attached to a negatively charged carboxylic ion. The head is called the hydrophilic (water-loving) end. It does not mix easily with oily substances and breaks apart easily in water. The tail of the molecule is called the hydrophobic (water-fearing) end. This end does not dissolve in water and bonds easily with oily or greasy substances, which are the biggest troublemakers when it comes to getting things clean. During the process of washing clothes, the hydrophilic end of the soap molecule decreases the surface tension of the water so that the water can be easily absorbed into the cloth. In other words, it makes it easier to wet the cloth. The hydrophobic end of the soap molecule helps pull dirt molecules out of the cloth by chemically bonding with them. Clusters of soap molecules, called micelles, form with their "water-loving" ends facing outward. Each micelle forms around a tiny particle of dirt, trapping it away from the clothing until the force of flowing water washes away both the dirt and the soap.

WHAT IS SODIUM HYDROXIDE?

Sodium hydroxide, also called caustic soda or lye, is a very dangerous chemical; just one drop can burn your skin. However, it's also a very important chemical with many uses beyond making soap. Its ability to corrode, or eat away at, other substances makes it the main ingredient in drain cleaners.

Did you ever wonder how paper is made from trees? Sodium hydroxide plays an important role in making paper. Cellulose, a fibrous plant material, is the basic material used to make most papers. The cellulose of woods, like birch and pine, must be dissolved and softened before it can be made into something called wood pulp, which is used to make paper. Cellulose can be a tough material and won't dissolve in water or alcohol, but a solution of sodium hydroxide is just strong enough to do the trick. It makes the cellulose fibers swell so that they are softer and easier to work with during the paper-making process.

Cellophane, the film often used to make packaging and clear tape, is also made from cellulose. In this case, the wood pulp is first treated with several chemicals to turn it into a thick fluid called viscose. The viscose is then pressed out into thin layers and treated with another sodium chemical called sodium sulfate (Na_2SO_4). Then it goes through another complex set of steps to become the clear film used to wrap packages.

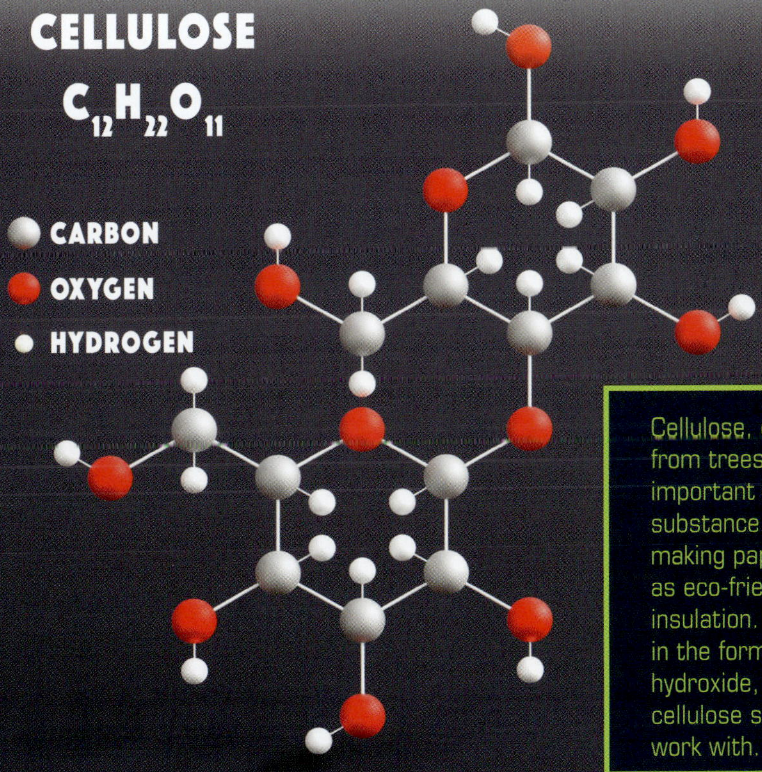

Cellulose, derived from trees, is a very important commercial substance. It's used for making paper, as well as eco-friendly cellulose insulation. It takes salt, in the form of sodium hydroxide, to make cellulose soft enough to work with.

Fireworks, which use sodium to produce bright colors, originated in ancient China. Although they're used now for entertainment, the first fireworks were probably signaling devices.

Olives are also treated with a sodium hydroxide solution before they are bottled or canned. By neutralizing a chemical called a glucoside, which makes the olives taste bitter, the sodium hydroxide solution helps make the olives taste better.

FIREWORKS!

Creating firework colors is a difficult job that combines art and science. The points of light ejected from fireworks (called stars) require an oxygen-producing chemical to sustain burning, a fuel, a color producer, and a substance called a binder to keep everything where it needs to be. There are two main mechanisms of color production in fireworks: incandescence and luminescence. Incandescence is light produced from heat. Heat causes a substance to become hot and glow, giving off red, orange, yellow, and white light as it becomes hotter. When the temperature of a firework is controlled, the glow of components can be manipulated to be the desired color at the proper time. Sodium is used in fireworks to give them a gold or yellow color.

Luminescence is light produced by energy sources other than heat. Sometimes luminescence is called "cold light" because it can occur at or below room temperature. To produce luminescence, energy is absorbed by an electron of an atom or molecule, causing it to become excited and move quickly. When the electron returns to a lower energy state, excess energy is released in the form of a light. The amount of energy present determines the color. Pure colors require pure ingredients. Even tiny amounts of sodium can affect the way a "star" appears.

ART IN THE SKY

The first fireworks were most likely used for military signals. Today, fireworks are dazzling displays used for entertainment. Fireworks are used to celebrate the Fourth of July and New Year's. In China they're used to celebrate the Lunar New Year, and in India, fireworks are part of Diwali celebrations.

The dazzling colors of fireworks are achieved by using certain elements.

COLOR	ELEMENTS
RED	STRONTIUM (SR), LITHIUM (LI)
ORANGE	CALCIUM (CA), STRONTIUM MIXED WITH SODIUM (NA)
GOLD	IRON (FE), SODIUM
YELLOW	SODIUM
WHITE	MAGNESIUM (MG), ALUMINUM (AL)
GREEN	BARIUM (BA) MIXED WITH CHLORINE (CL)
BLUE	COPPER (CU) MIXED WITH CHLORINE
PURPLE	STRONTIUM MIXED WITH COPPER
SILVER	ALUMINUM, TITANIUM (TI), ZIRCONIUM (ZR), OR MAGNESIUM

LIGHTING UP THE NIGHT

Low-pressure sodium lamps were invented in the 1920s. Commercial production began in 1932. They produce yellow light and are still common in Europe. High-pressure sodium lamps were introduced in the 1960s. They're used worldwide for streetlights. Sodium lamps produce light by burning neon and sodium gas. They use all their electrical voltage to produce light in wavelengths visible to the human eye. They're more efficient than incandescent lamps, which produce light in all wavelengths, using electrical current to produce light that isn't visible to humans.

Low-pressure sodium vapor lamps produce a distinctive yellow light. They're more commonly used in Europe for outdoor lighting.

CHAPTER 5

SODIUM AND BIOLOGY

The human body requires sodium to function. Sodium is an electrolyte—a mineral that carries an electrical charge. Salt helps nerves and muscles function properly. It also helps regulate the amount of fluid in the body. Generally, people get the sodium they need (around 500 mg) from the food they eat—in fact, most people eat more salt than their body requires. Your kidneys regulate the amount of sodium in your body by excreting the excess in sweat and urine.

A certain amount of sodium is necessary for healthy blood pressure. However, too much salt (hypernatremia) can cause hypertension, or high blood pressure. Too little salt (hyponatremia) isn't good either. You may have noticed if you're very active on a hot, sunny day and sweat a great deal, you don't feel well. This is because sweat isn't just water—it contains minerals, including salt, that your body needs.

Eating potato chips, which contain salt and potassium, or drinking a sports drink that replenishes your electrolytes can make you feel better if you're sweating excessively.

IT'S ELECTRIC!

Table salt is formed by ionic bonds between sodium and chlorine. When salt is dissolved in water, the bonds break, releasing positively charged sodium ions and negatively charged chlorine ions into the water. This liquid will conduct electricity. Fluids and ions within the human body that perform this way are called electrolytes. Sodium, calcium, potassium, and magnesium all work in the human body as electrolytes.

Brightly colored sports drinks high in electrolytes are available everywhere. But are they good for you? It depends. If you're sweating excessively due to overexertion, overheating, or fever, they can be very helpful. But if you're not losing electrolytes, too much can make you sick.

Electrolytes in the blood help the body function properly. For example, they play an important role in movement. In the nervous system, they send tiny electrical signals to nerve cells. When the nerve cells receive this signal, they release a chemical called acetylcholine, which travels to a nearby muscle fiber and prepares it to accept the sodium and potassium ions coming its way. As they travel, these ions signal calcium ions to help set up a series of chemical reactions that cause your muscles to move.

Water continually circulates through the body, especially in plasma, the liquid part of the blood. A large part of this flow happens because of sodium chloride.

Within the body, sodium ions are needed in high concentrations outside of cells and in low concentrations within cells. As blood moves through the organs of the body, substances move in and out of the blood by osmosis. For example, whenever you drink a glass of water, far more of that water passes through your kidneys than is released as urine. As the blood carries waste products from the body into the kidneys, water flows through the kidneys and back into the blood through osmosis. As this happens, the waste substances gather in one place to exit from your body as urine. Some salt and water are always eliminated along with the waste, which is why it's important that your diet includes salt (in moderation) and water to replenish what is lost.

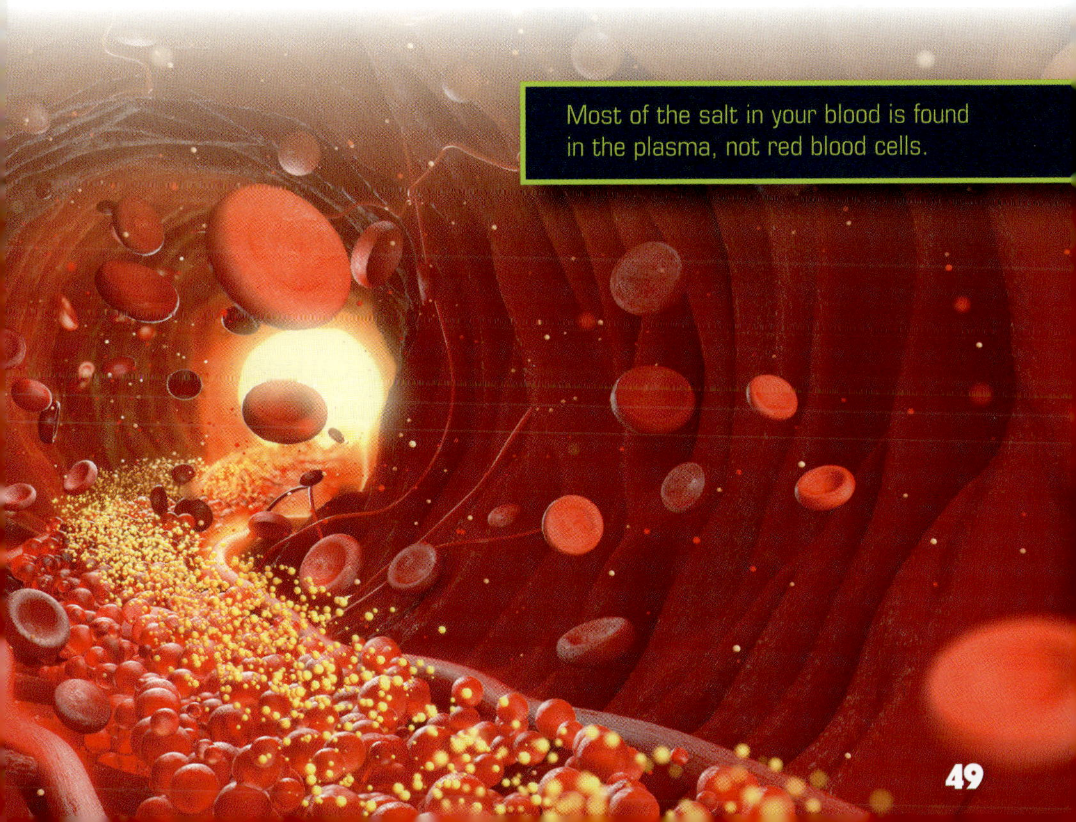

Most of the salt in your blood is found in the plasma, not red blood cells.

Baking soda is one of the most useful substances in your kitchen. It can be used for cooking, cleaning, and even putting out grease fires.

BAKING SODA

In the 1860s, Belgian chemist Ernest Solvay (1838–1922) discovered a way to make sodium bicarbonate ($NaHCO_3$), the chemical we call baking soda. First, Solvay heated calcium carbonate, or limestone—a light-colored rock often used in building. When heated or burned, limestone breaks down into calcium oxide (CaO), also known as lime, and carbon dioxide (CO_2). The carbon dioxide was mixed with a solution of sodium chloride and ammonia (NaH_3) to produce sodium bicarbonate and ammonium chloride (NaH_4Cl).

Sodium bicarbonate has many uses. When mixed with flour and water to make dough, baking soda becomes carbon dioxide. As this happens, bubbles are formed in the dough that make it lighter. This process gives bread, biscuits, and other baked foods their fluffy texture.

Baking soda is useful in treating heartburn, a painful condition that occurs when the stomach produces too much acid. The baking soda neutralizes the acid but often produces carbon dioxide gas as a side effect. This gas can cause pressure on the stomach, making a person feel very uncomfortable.

Have you ever noticed that many toothpastes on the market today contain baking soda? This is because baking soda is also a gentle abrasive that can make teeth whiter, while naturally absorbing odors. This odor-absorbing quality is also the reason why many people keep a box of baking soda in the refrigerator and in the closet near smelly shoes.

Tarnish can be removed from silver using a solution of vinegar, baking soda, and boiling water. The secret ingredient is a piece of aluminum foil. The chemical reaction between the ingredients transfers the tarnish from the silver to the aluminum.

Acetic acid (vinegar) releases carbon dioxide when it reacts with sodium bicarbonate. This is visible as fizzy bubbles!

Baking soda can clean so many different things and remove odors because it alters the pH level of things it mixes with. The letters "pH" indicate the amount of hydrogen ions a substance contains. More hydrogen ions equal greater acidity, indicated by a lower pH number (zero to six). Fewer hydrogen ions mean something is more basic or alkaline. This is indicated by a higher pH number (up to 14). Pure water is neutral and has a pH of seven.

When baking soda mixes with something acidic, it reacts, releasing water, salt, and carbon dioxide. This mixture can soothe stomach acid, clean many different surfaces, and even put out fire. (The carbon dioxide deprives flames of oxygen, smothering them.)

As we've seen, sodium and its many compounds are all around us. Salt has been an important part of human civilization for thousands of years. Ancient Egyptians used a form of sodium as part of the mummification process. Sodium is used in the manufacture of soap and drain cleaner. It makes food taste better and fireworks sparkle. It helps plants grow and keeps our bodies healthy. We literally can't live without it!

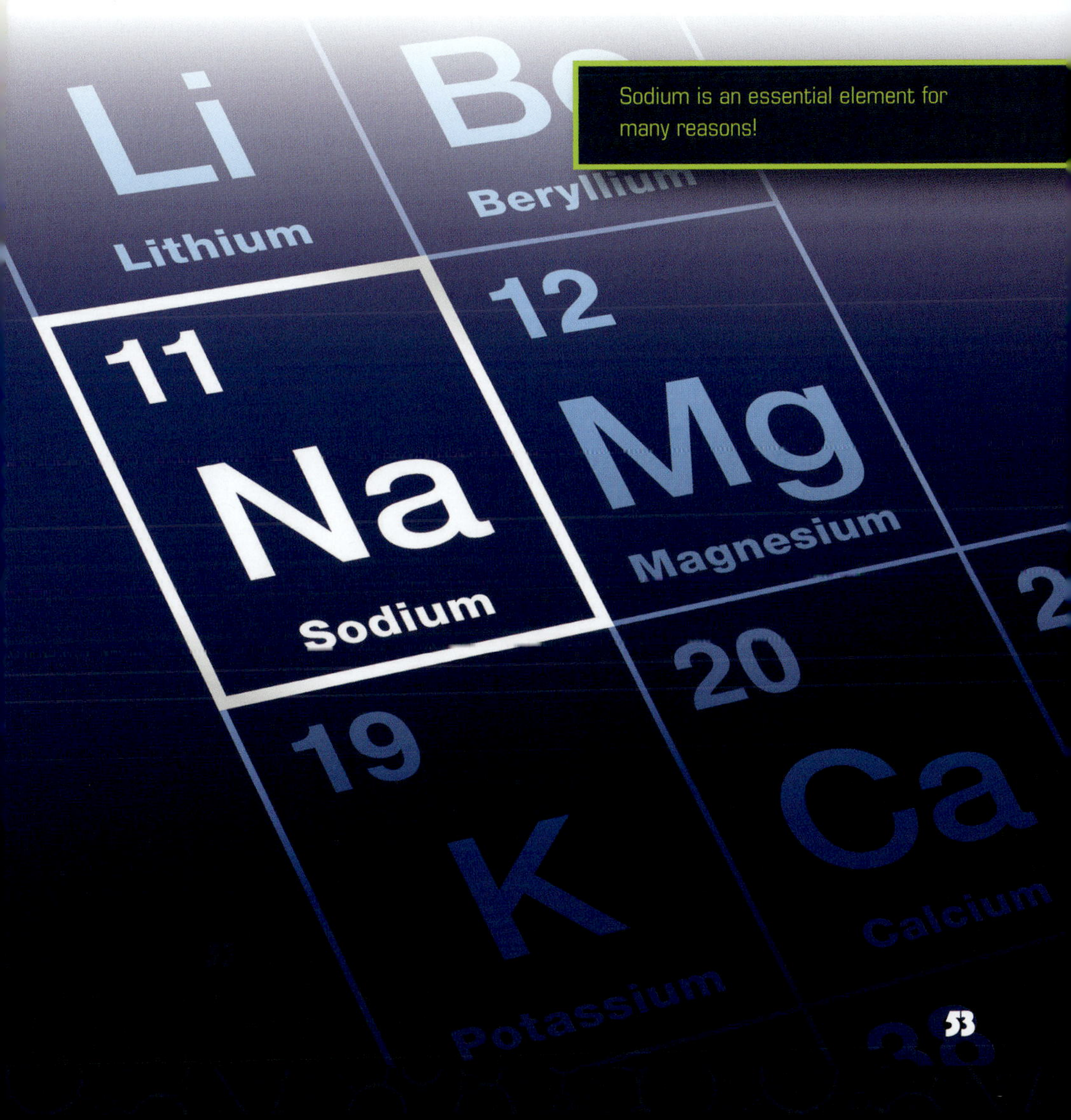

Sodium is an essential element for many reasons!

THE PERIODIC TABLE OF ELEMENTS

By studying the periodic table, we can learn a lot about elements such as sodium. Learning the periodic table is an important part of the science of chemistry.

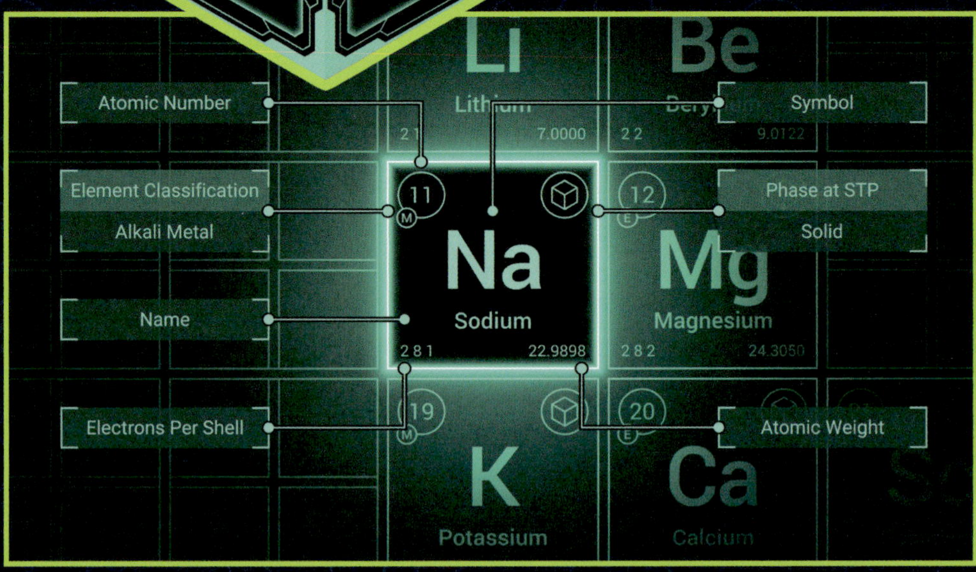

PERIODIC TABLE OF ELEMENTS

1 **H** Hydrogen 1.008																	2 **He** Helium 4.0026
3 **Li** Lithium 6.94	4 **Be** Beryllium 9.0122											5 **B** Boron 10.81	6 **C** Carbon 12.011	7 **N** Nitrogen 14.007	8 **O** Oxygen 15.999	9 **F** Fluorine 18.998	10 **Ne** Neon 20.180
11 **Na** Sodium 22.990	12 **Mg** Magnesium 24.305											13 **Al** Aluminium 26.982	14 **Si** Silicon 28.085	15 **P** Phosphorus 30.974	16 **S** Sulfur 32.06	17 **Cl** Chlorine 35.45	18 **Ar** Argon 39.948
19 **K** Potassium 39.098	20 **Ca** Calcium 40.078	21 **Sc** Scandium 44.956	22 **Ti** Titanium 47.867	23 **V** Vanadium 50.942	24 **Cr** Chromium 51.996	25 **Mn** Manganese 54.938	26 **Fe** Iron 55.845	27 **Co** Cobalt 58.933	28 **Ni** Nickel 58.693	29 **Cu** Copper 63.546	30 **Zn** Zinc 65.38	31 **Ga** Gallium 69.723	32 **Ge** Germanium 72.630	33 **As** Arsenic 74.922	34 **Se** Selenium 78.971	35 **Br** Bromine 79.904	36 **Kr** Krypton 83.798
37 **Rb** Rubidium 85.468	38 **Sr** Strontium 87.62	39 **Y** Yttrium 88.906	40 **Zr** Zirconium 91.224	41 **Nb** Niobium 92.906	42 **Mo** Molybdenum 95.95	43 **Tc** Technetium (98)	44 **Ru** Ruthenium 101.07	45 **Rh** Rhodium 102.91	46 **Pd** Palladium 106.42	47 **Ag** Silver 107.87	48 **Cd** Cadmium 112.41	49 **In** Indium 114.82	50 **Sn** Tin 118.71	51 **Sb** Antimony 121.76	52 **Te** Tellurium 127.60	53 **I** Iodine 126.90	54 **Xe** Xenon 131.29
55 **Cs** Caesium 132.91	56 **Ba** Barium 137.33	57-71	72 **Hf** Hafnium 178.49	73 **Ta** Tantalum 180.95	74 **W** Tungsten 183.84	75 **Re** Rhenium 186.21	76 **Os** Osmium 190.23	77 **Ir** Iridium 192.22	78 **Pt** Platinum 195.08	79 **Au** Gold 196.97	80 **Hg** Mercury 200.59	81 **Tl** Thallium 204.38	82 **Pb** Lead 207.2	83 **Bi** Bismuth 208.98	84 **Po** Polonium (209)	85 **At** Astatine (210)	86 **Rn** Radon (222)
87 **Fr** Francium (223)	88 **Ra** Radium (226)	89-103	104 **Rf** Rutherfordium (267)	105 **Db** Dubnium (268)	106 **Sg** Seaborgium (269)	107 **Bh** Bohrium (270)	108 **Hs** Hassium (277)	109 **Mt** Meitnerium (278)	110 **Ds** Darmstadtium (281)	111 **Rg** Roentgenium (282)	112 **Cn** Copernicium (285)	113 **Nh** Nihonium (286)	114 **Fl** Flerovium (289)	115 **Mc** Moscovium (290)	116 **Lv** Livermorium (293)	117 **Ts** Tennessine (294)	118 **Og** Oganesson (294)
			57 **La** Lanthanum 138.91	58 **Ce** Cerium 140.12	59 **Pr** Praseodymium 140.91	60 **Nd** Neodymium 144.24	61 **Pm** Promethium (145)	62 **Sm** Samarium 150.36	63 **Eu** Europium 151.96	64 **Gd** Gadolinium 157.25	65 **Tb** Terbium 158.93	66 **Dy** Dysprosium 162.50	67 **Ho** Holmium 164.93	68 **Er** Erbium 167.26	69 **Tm** Thulium 168.93	70 **Yb** Ytterbium 173.05	71 **Lu** Lutetium 174.97
			89 **Ac** Actinium (227)	90 **Th** Thorium 232.04	91 **Pa** Frotactinium 231.04	92 **U** Uranium 238.03	93 **Np** Neptunium (237)	94 **Pu** Plutonium (244)	95 **Am** Americium (243)	96 **Cm** Curium (247)	97 **Bk** Berkelium (247)	98 **Cf** Californium (251)	99 **Es** Einsteinium (252)	100 **Fm** Fermium (257)	101 **Md** Mendelevium (258)	102 **No** Nobelium (259)	103 **Lr** Lawrencium (266)

- Alkali metals
- Alkaline earth metals
- Transition metals
- Metal
- Metalloids
- Nonmetals
- Halogen
- Noble gases
- Lanthanoids
- Actinoids

- C Solid
- Hg Liquid
- H Gas
- Rf Unknown

GLOSSARY

acid: A reactive substance that accepts electrons or donates protons; usually has a sour, sharp, or biting taste.

alkali: A soluble salt obtained from the ashes of plants.

aquifer: An underground bed or layer of water.

atom: The smallest part of an element.

base: A water-soluble and bitter-tasting compound.

chemical reaction: A change in which one kind of matter is turned into another kind of matter.

electrolysis: A process that produces chemical changes by the passage of an electric current.

electron: A negatively charged particle found outside the nucleus of an atom.

fibrous: Having or resembling fiber or stringy tissue.

inert: Lacking a usual or anticipated chemical or biological action.

ion: A particle that is positively or negatively charged; an atom that has lost or gained one or more electrons.

ionic bond: A chemical bond formed between oppositely charged species because of their mutual electrostatic attraction.

isotope: An atom of a chemical element with the same atomic number and nearly identical chemical behavior but with a different atomic weight and different physical properties.

matter: Material substance that occupies space.

molecule: The smallest bit of matter before it gets broken down into its basic parts, or atoms.

neutron: A particle within the nucleus of an atom that contains no charge; found in the nucleus of all elements except hydrogen.
nucleus: The positively charged center portion of an atom.
phase: The state of something. The three fundamental phases of matter are solid, liquid, and gas.
proton: A positively charged particle within the nucleus of an atom. The number of protons and electrons are almost always equal in an atom.
radioactive: Of, caused by, or exhibiting radioactivity, which is the ability to spontaneously produce energy.
stable: Not spontaneously radioactive.

FOR MORE INFORMATION

AMERICAN CHEMICAL SOCIETY

1155 16th Street NW
Washington, DC 20036
(800) 227-5558
Website: www.acs.org
Facebook: @AmericanChemicalSociety
Instagram and X: @AmerChemSociety
ACS is a global professional community of chemists.

CHEMICAL INSTITUTE OF CANADA (CIC)

90-2420 Bank Street
Ottawa, Ontario, Canada
K1V 8S1
(613) 232-6252
Website: www.cheminst.ca/
X: @CIC_ChemInst
The CIC facilitates the research and cooperation of chemists and chemical engineers in Canada.

INTERNATIONAL UNION OF PURE AND APPLIED CHEMISTRY (IUPAC)

IUPAC Secretariat
79 T.W. Alexander Drive
Research Commons Building 4501, Suite 190
Research Triangle Park, NC 27709, USA
(919) 485 8700
Website: iupac.org/
Facebook: @iupac.org
X: @iupac
IUPAC is the global chemistry organization that maintains the periodic table of elements.

INTERNATIONAL UNION OF PURE AND APPLIED PHYSICS (IUPAP)

Fondazione Internazionale Trieste per il Progresso e la Libertà delle Scienze
c/o ICTP
Strada Costiera 11
34151 Trieste
Website: iupap.org/
Facebook: @iupap
X: @iupap_physics
IUPAP's mission is to assist in the global advancement of physics and facilitate international scientific cooperation.

NATIONAL GEOGRAPHIC SOCIETY

1145 Seventeenth Street NW
Washington, DC 20036
Website: www.nationalgeographic.com
Facebook, Instagram, and X: @natgeo
This group sponsors scientific expeditions to explore the natural world. Its documentary films and magazine are world renowned.

THE SMITHSONIAN INSTITUTION

Smithsonian Information
Ripley Center, MRC 010
PO Box 37012
Washington, D.C. 20013-7012
202-633-1000
Website: www.si.edu/
Facebook and Instagram: @smithsonian
The Smithsonian is the world's largest research, education, and museum complex. It encompasses numerous museums in Washington, D.C., and New York City, as well as a world-class zoo.

FOR FURTHER READING

Congdon, Lisa. *The Illustrated Encyclopedia of the Elements: The Powers, Uses, and Histories of Every Atom in the Universe*. San Francisco, CA: Chronicle Books, LLC, 2021.

DK. *Super Simple Chemistry: The Ultimate Bitesize Study Guide*. New York, NY: DK Publishing, 2020.

McHenry, Ellen Johnston. *The Chemical Elements Coloring & Activity Book: A Fun and Interactive Guide to the World of Chemistry*. Independently Published, 2021.

Silver, Donald and Patricia Wynne. *My First Book About Chemistry*. Mineola, NY: Dover Publications, 2020.

Thomas, Isabel, and Sara Gillingham. *Exploring the Elements: A Complete Guide to the Periodic Table*. New York, NY: Phaidon Press, Inc. 2021.

Zovinka, Edward P., and Rose A. Clark. *A Kids' Guide to the Periodic Table: Everything You Need to Know About the Elements*. Emeryville, CA: Rockridge Press, 2020.

INDEX

ABOUT THE AUTHOR

KATHLEEN A. KLATTE is the author of many nonfiction books for children and teens. Topics she has written about range from animals and nature to constitutional law to unusual career choices. She lives in New York with one cat and far too many books and Legos.

PHOTO CREDITS

Cover, p. 1 remotevfx.com/Shutterstock.com; Cover, pp. 1, 3-64 Oksancia/Shutterstock.com; Cover, pp. 1, 3-64 pluie_r/Shutterstock.com; p. 5 ODMa/Shutterstock.com; p. 7 Firsova Natalia/Shutterstock.com; p. 8 Mishainik/Shutterstock.com; pp. 9, 22, 39 New Africa/Shutterstock.com; p. 11 historiasperiodicas/Shutterstock.com; p. 12 Srikanth_K/Shutterstock.com; p. 13 Sergey Demo SVDPhoto/Shutterstock.com; p. 14 Designua/Shutterstock.com; p. 15 Thais29/Shutterstock.com; p. 17 Absolute/Shutterstock.com; p. 19 OSweetNature/Shutterstock.com; p. 20 Heiko Kueverling/Shutterstock.com; p. 23 Juan Roballo/Shutterstock.com; p. 25 stamford/Shutterstock.com; p. 26 Jag_cz/Shutterstock.com; p. 27 NayaDadara/Shutterstock.com; p. 28 Veyselcelikdemir/Shutterstock.com; p. 29 kavram/Shutterstock.com; p. 30 Blueee77/Shutterstock.com; p. 31 FAIZ IBRA/Shutterstock.com; p. 32 S.Narongrit99/Shutterstock.com; p. 33 Alf Ribeiro/Shutterstock.com; p. 35 pilipphoto/Shutterstock.com; p. 37 PeopleImages.com - Yuri A/Shutterstock.com; p. 38 Love Employee/Shutterstock.com; p. 41 SergeiShimanovich/Shutterstock.com; p. 42 Oncewaslost1/Shutterstock.com; p. 45 Barry Paterson/Shutterstock.com; p. 47 Zenzeta/Shutterstock.com; p. 48 Halfpoint/Shutterstock.com; p. 49 VFXartist/Shutterstock.com; p. 50 HalynaRom/Shutterstock.com; p. 51 KrimKate/Shutterstock.com; p.52 EKramar/Shutterstock.com; p. 53 Peter Hermes Furian/Shutterstock.com; p. 54 ShannonChocolate/Shutterstock.com; p. 55 View_From_My_Lens/Shutterstock.com.